ALIEN SITES

by Jenny Mason

12

STORY LIBRARY

MORE TO EXPLORE

www.12StoryLibrary.com

12-Story Library is an imprint of Bookstaves.

Developed and produced for 12-Story Library by Focus Strategic Communications Inc.

Library of Congress Cataloging-in-Publication Data
Names: Mason, Jenny (Children's author), author.
Title: Alien sites / by Jenny Mason.
Description: Mankato, Minnesota : 12-Story Library, [2022] | Series: Aliens! |
Includes bibliographical references and index. | Audience: Ages 10–13 | Audience: Grades 4–6
Identifiers: LCCN 2020014852 (print) | LCCN 2020014853 (ebook) | ISBN 9781632359322 (library binding) | ISBN 9781632359674 (paperback) | ISBN 9781645820987 (pdf)
Subjects: LCSH: Extraterrestrial beings—Juvenile literature. | Unidentified flying objects—Sightings and encounters—Juvenile literature. | Historic sites—Juvenile literature.
Classification: LCC QB54 .M37 2022 (print) | LCC QB54 (ebook) | DDC 001.942—dc23
LC record available at https://lccn.loc.gov/2020014852
LC ebook record available at https://lccn.loc.gov/2020014853

Photographs ©: sdecoret/Shutterstock.com, cover; Franck Fotos/Alamy, 1; Daniel Prudek/Shutterstock.com, 4; Kursor/Shutterstock.com, 5; SL-Photography/Shutterstock.com, 5; ZUMA Press, Inc./Alamy, 6; Manifesta Prod/YouTube.com, 6; To The Stars Media/YouTube.com, 7, USA Today/YouTube.com, 7; Dean Clarke/Shutterstock.com, 8; ehrlif/Shutterstock.com, 8; Digital Storm/Shutterstock.com, 9; Dudeanatortron/CC4.0, 9; R s/YouTube.com, 10; Troy Vovchek/Shutterstock.com, 10; andrea crisante/Shutterstock.com, 10; XX, 11; KTNV Channel 13 Las Vegas/YouTube.com, 11; Nawabtanweer/CC4.0, 12; Soban/CC3.0, 13; Shashank Agarwal/Shutterstock.com, 13; Sheri Swailes/Shutterstock.com, 14; MWaits/Shutterstock.com, 14; photoBeard/Shutterstock.com, 15; Grossinger/Shutterstock.com, 15; XX, 16; DenversAirport/YouTube.com, 17; Ambient Ideas/Shutterstock.com, 17; Mark Green/Shutterstock.com, 18; Svineyard/Shutterstock.com, 19; IanDagnall Computing/Alamy, 19; XX, 20; IgorZh/Shutterstock.com, 20; nraliessi/CC2.0, 21; NASA/Ames Research Center/W. Stenzel/D. Rutter, 22; NASA/Terry Virts, 23; Serjio74/Shutterstock.com, 23; ROGER NORMAN/Alamy, 24; Damian Pankowiec/Shutterstock.com, 24; Skyring/CC3.0, 25; Franck Fotos/Alamy, 26; XX, 26; James R. Poston/Shutterstock.com, 27; KeremGogus/Shutterstock.com, 28; ZUMA Press, Inc./Alamy, 28; Aleksandra Suzi/Shutterstock.com, 29

About the Cover

An alien flying saucer makes a visit.

Access free, up-to-date content on this topic plus a full digital version of this book. Scan the QR code on page 31 or use your school's login at 12StoryLibrary.com.

Table of Contents

Nazca Lines: Landing Strips for Spaceships?

Aerial view of hands and a tree.

The Nazca Lines are ruts, or geoglyphs, cut deep into the dry desert soil. The ruts stretch in straight lines for 30 miles (50 km) or more. Some connect to hundreds of drawings that include plants, animals, and imaginary monsters.

The huge artwork is located 200 miles (322 km) south of Lima, Peru's capital. Researchers estimate that the images appeared almost 2,500 years ago. The prehistoric Nazca people carved them. Scientists aren't sure why.

Did they guide people to a sacred hill? Did they channel

Some think the strange lines guided alien spaceships.

water? Were they landing strips for alien spaceships? Why else carve more than 800 straight long lines?

THINK ABOUT IT

Why did the Nazca carve the geoglyphs? What other reasons can you think of?

1,200

Distance across in feet (365 m) of the largest Nazca drawings

- The Nazca drew almost 400 shapes and pictures.
- The drawings include a monkey, a hummingbird, a spider, a jaguar, and a freckled monster sticking out its tongue.
- Toribio Mejia Xesspe, a Peruvian archaeologist, discovered the lines in 1927.

2

To the Stars Academy: Alien School

Tom DeLonge was a member of the band Angels & Airwaves.

Musician Tom DeLonge had a problem. He headed a famous rock and roll band, but he loved UFO-hunting more. Finally he quit the band. In 2014, he created To the Stars Academy of Arts & Science in 2014. The company is a school about aliens.

The Academy attracts experts from NASA, national security, and the military. It unites

Pilots on F-18 aircraft reported sightings of UFOs.

CURIOUS, NOT UNBELIEVABLE

Luis Elizondo led a top-secret UFO search program for the US government. He was curious about life in the universe. He left the job because people didn't believe him. Instead, Elizondo began working for the Academy.

scientists, artists, and entertainers. They spread information about life in the universe.

To do this, the Academy makes movies and documentaries. They publish books and research space mysteries.

They also give money for space-travel projects, like warp drives. These engines could make spaceships faster.

2017

Year the Academy released three UFO videos from the US Navy

- The videos show encounters between military aircraft and UFOs.
- They made a UFO documentary for the History Channel.
- The Academy created an online UFO database.

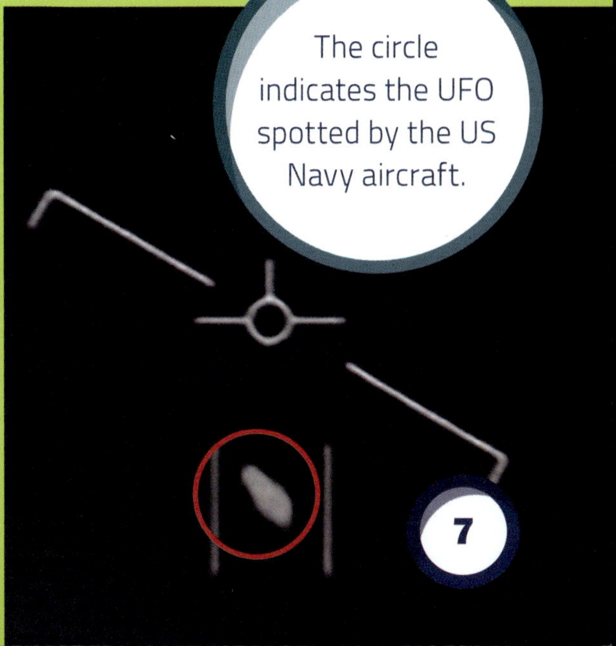

The circle indicates the UFO spotted by the US Navy aircraft.

3

Area 51: Flying Under the Radar

A group of buildings in Nevada's desert didn't exist until 2013. That year, the government finally confessed. A hidden facility was located there. It was called Area 51.

People gossiped about Area 51. They said the military stored an alien spaceship there. There could be alien remains, too. Maybe the military was using alien technology.

The rumors were partly correct. Pilots learned to fly stealth jets there. The U-2 spy plane. The OXCART jet. The F-117 Nighthawk stealth fighter.

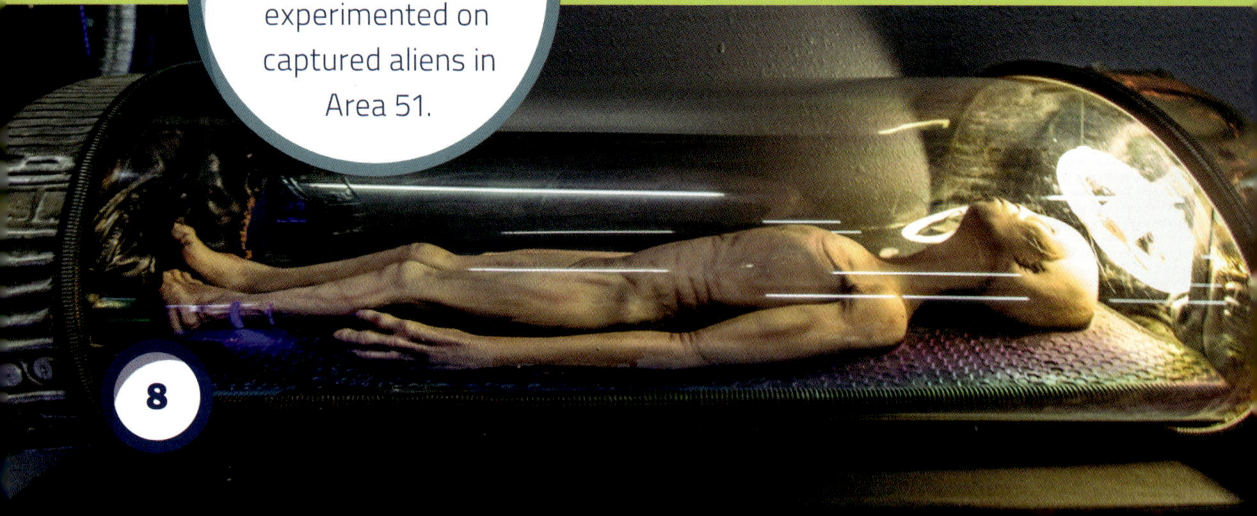

Some people believe that scientists experimented on captured aliens in Area 51.

The F-117 Nighthawk was one of the aircraft flown at Area 51.

The SR-71 Blackbird. Those jets were flown at Area 51. Those oddly shaped, super-fast planes resemble UFOs.

LIAR, LIAR

In 1989, Robert Lazar talked to the press. He said he had worked on alien spaceships at Area 51. His stories fueled wild conspiracies. After much investigation, reporters found Lazar had lied.

70,000

Altitude in feet (21,000 m) that the U-2 spy plane reached

- Some Area 51 aircraft reached speeds of 2,000 miles per hour (3,200 km/h).
- Area 51 had other code names like, "Dreamland" and "Paradise Ranch."
- Astronauts accidently photographed Area 51 from space in 1974.

4

Storm Area 51: A Failure to Launch

Matty Roberts.

Matty Roberts only meant to be funny. In 2019 the video gamer created a fake Facebook event. He called it Storm Area 51, asking people to raid Area 51. Area 51 is a US Air Force facility in the Nevada desert.

Some people believed alien spacecraft were stored in Area 51.

$200,000

Amount spent to clean up after Storm Area 51

- The nearby town of Rachel has no gas station, streetlamps, or grocery store.
- The event left a lot of litter, which the town paid to remove.
- The Little A'Le'Inn in Rachel has appeared in several TV shows, a movie, and a video game.

According to rumors, the military stored wrecked alien ships and alien bodies inside. Roberts did not really plan to raid the base. Still, two million people responded. As a result, the FBI questioned Roberts.

In the end, 3,000 people gathered. The nearby small town of Rachel hosted an alien festival. About 150 people went to Area 51. They did not tangle with the armed guards.

Fewer people came to the Storm Area 51 event than expected.

5

Mohenjo-Daro: Alien Warzone

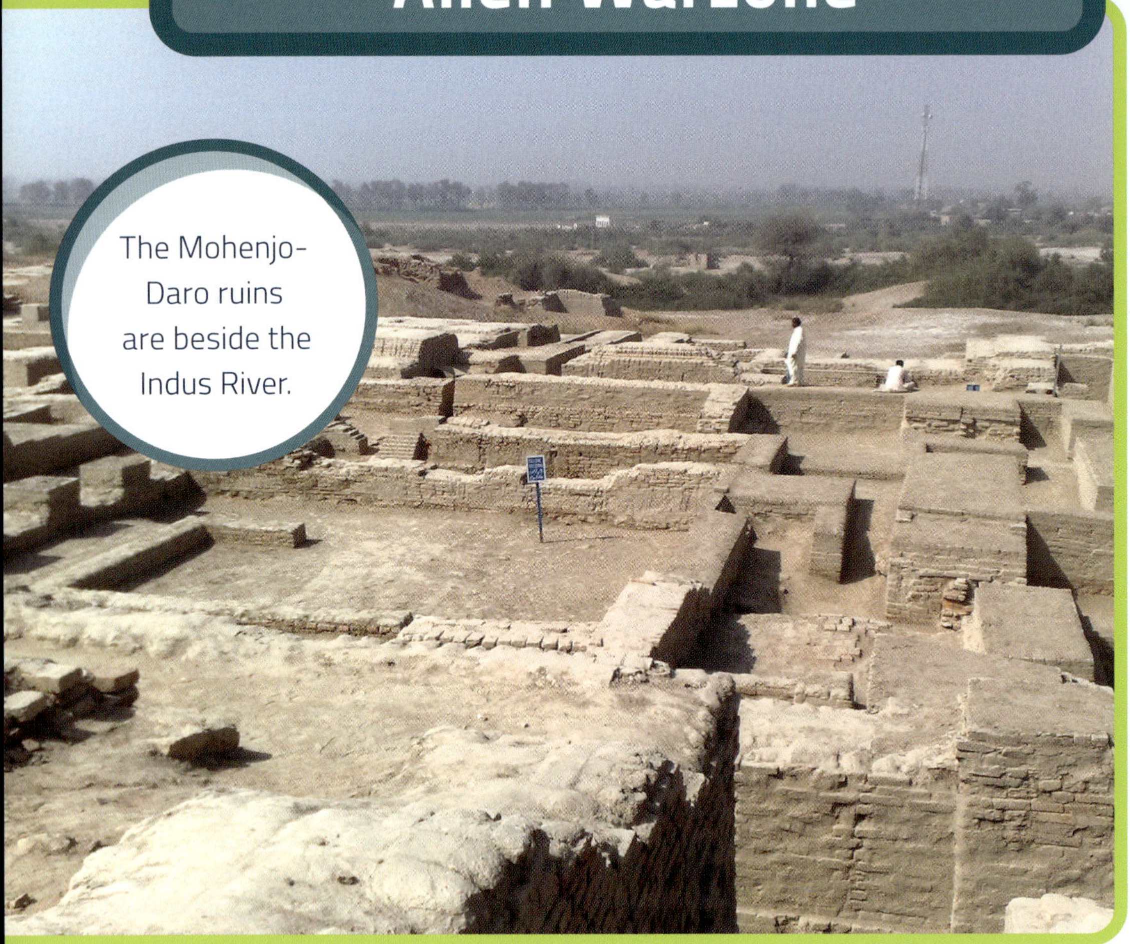

The Mohenjo-Daro ruins are beside the Indus River.

In southern Pakistan, there is an ancient abandoned city called Mohenjo-Daro. Between 2,600 and 1,500 BCE, Mohenjo-Daro grew. It was one of the world's largest, most advanced cities.

Many skeletons have been found in the city. One large area appears to have been blasted. It looks like melted glass. To some, the destruction resembles a nuclear bomb blast.

Some think aliens fought an epic battle in Mohenjo-Daro.

After much study, scientists debunked this theory. The skeletons were buried over many years. The melted glass is actually pottery and the remains of ancient kilns.

750

Total acres (300 hectares) covered by the Mohenjo-Daro ruins

- Mohenjo-Daro translates to *Mound of the Dead*. The original name of the city is unknown.
- The city was divided into rectangular blocks.
- As many as 40,000 people lived in Mohenjo-Daro.

CARRIED AWAY

Imagination is one of humankind's most powerful skills. It can also muddle science. For example, an iron loop with decorative pieces looked like a crown to archaeologists. It turns out it was the handle and top of an ancient bucket.

6

Roswell, New Mexico: When Facts Balloon

An alien spaceship display at the International UFO Museum and Research Center in Roswell, New Mexico.

In 1947, Roswell, New Mexico, grabbed headlines. Thirty miles (50 km) north of town, a mystery arose. A rancher rode across the land. He found a long gash in the ground. It led to a wrecked aircraft.

News reports claimed it was a flying saucer. The military said it was just a weather balloon. They quickly cleaned the area. Then, local rumors ballooned. Some people swore they saw alien bodies. Decades later, one man videotaped a fake medical exam of alien bodies. People thought the footage was real.

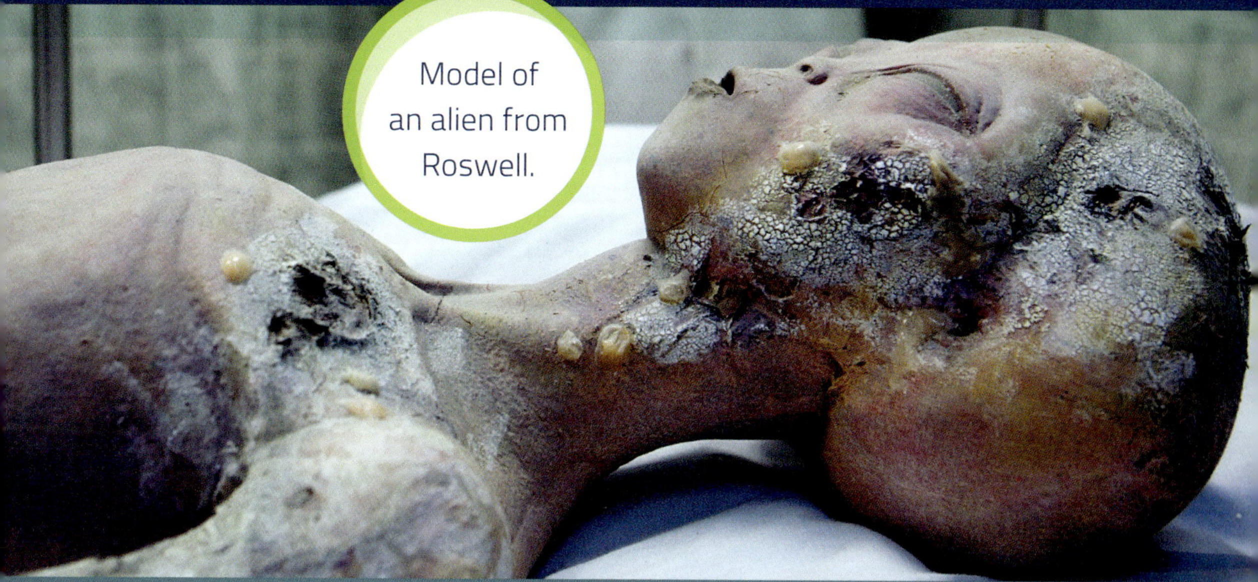

Model of an alien from Roswell.

THINK ABOUT IT

Have you ever heard facts exaggerated by rumors? How did information change over time?

A balloon did crash that day. It was from a secret spy mission. Government scientists put devices in hot air balloons to listen for enemy bomb tests.

$6 million

Amount tourists spent in Roswell during the 2017 annual UFO Festival

- Roswell enjoys alien-mania with alien streetlamps, alien restaurants, and an International UFO Museum and Research Center.
- The museum welcomed 200,000 visitors in 2016.
- Roswell is the fifth largest city in New Mexico with 48,000 residents.

7 Denver International Airport: Runway Rumors

Gossip has always shadowed the Denver International Airport (DIA, or DEN). Of course, it is one of the world's strangest travel hubs. For example, it has a 32-foot (10 m) blue mustang statue with glowing red eyes. It is nicknamed Blucifer. Some people say it's cursed. A piece of the 9,000-pound (4,100 kg) statue fell and crushed the sculptor.

In 2018, DIA began a $770 million renovation. This project sparked more rumors. It was probably because of the construction barriers. They invite people to guess what is being built. Zombie cat lairs? Area 52? Gargoyle breeding grounds?

53

Area in square miles (137 sq km) DIA covers above ground

- It is one of the biggest airports in the world.
- It has an annual budget of $2 billion.
- The main terminal is a long strip of white tents. They resemble the snowy tips of Colorado's Rocky Mountains.

A bizarre statue chats with passengers. It has a moving face and mouth. It is a robot in disguise.

8

Puma Punku: Alien Jigsaw Puzzle

Puma Punku was once a mighty temple. Giant stone blocks formed its walls. Some of them weigh over 100 tons (91 metric tons). The notched rocks fit like puzzle pieces. Today, the temple is mostly rubble.

Some scholars thought aliens built the structure. Only advanced technology could have moved the massive stones. Advanced tools, like lasers, must have notched them.

In fact, simple ropes moved the stones. The builders used metal tools for the notches.

Puma Punku is a UNESCO World Heritage Site in Bolivia.

Researchers made other mistakes. For example, they misjudged the temple's age.

A BRILLIANT BLUNDER

Albert Einstein made a famous mistake. In the 1900s, great minds agreed the cosmos was static. Einstein's math contradicted this belief, so he changed his equations. The decision was a mistake. The universe is expanding. Unfortunately, aliens are even less likely to find and visit Earth in such a universe.

2

Number of football fields that would fit into Puma Punku

- The Tiwanaku people built the temple. They have since disappeared.
- Centuries later, the Inca found the site. They thought gods had made it.
- High-tech Tiwanaku farming supported 400,000 people in the area.

Emperor Atahualpa was the last Inca ruler.

9

Ovnipuerto: Whispers from Above

Werner Jaisli built a UFO airport in Argentina. Reporters wanted to know why. He said aliens ordered him to. In 2008, Jaisli came to Argentina from Switzerland. One night, two glowing objects hovered near him.

The aliens did not speak. They gave instructions telepathically, or through their thoughts. Jaisli traveled to the small town of Cachi. There, he formed a

Werner Jaisli.

Aerial view of Ovnipuerto with landing strip.

massive 36-pointed star using bright white stones. Inside that, he piled brown stones into a smaller star. He called the airport Ovnipuerto, which is Spanish for UFO port.

The stars would guide aliens. After the airport was completed, Jaisli disappeared. No one knows where he is today.

150

Distance across in feet (46 m) of the largest star

- Ovnipuerto stars can be seen from 7,200 feet (2,200 m) above.
- Many tourists, ufologists, and other UFO fans visit Ovnipuerto every year.
- Some Cachi locals have recorded UFO sightings, but no aliens have landed...so far.

THINK ABOUT IT

Extreme fatigue can cause hallucinations, or illusions mixing reality and dreams. How might researchers test whether a UFO sighting is real or a hallucination?

10

Welcome Home: Earth's Alien Worlds

UFO and alien sites on Earth inspire scientists to search for life in outer space. In recent years, astronomers discovered thousands of planets (exoplanets). Each one is strange and wonderful. Aliens have not been discovered on these planets yet. Perhaps climate change or big wars drove them away. UFO supporters claim that aliens seek Earth as a new home. Luckily, Earth has bizarre alien worlds already.

Aliens might feel at home in the Danakil Depression in

The Kepler space telescope is shown surrounded by exoplanets. Scientists believe there are more exoplanets than stars.

Ethiopia. This sunken volcano has acid ponds. It is splattered in neon chemical colors. Mineral chimneys spit out rocks. Salar de Uyuni in Bolivia is the world's largest salt desert. It is visible from space. Maybe it has attracted aliens.

FAMOUS ALIEN DESERT

Moviemakers use places on Earth to stand in for alien planets. The Wadi Rum desert in Jordan often plays the role of Mars in many movies.

Moon Valley in Atacama desert.

15 million

Age in years of Chile's Atacama Desert, the oldest on Earth

- Paprika-red sands and toothy salt slabs create a very alien world.
- NASA tests new Mars rovers in Atacama.
- The high desert is also alien-hunting headquarters. Many telescopes there search for life in space.

11

Wycliffe Well: Aliens in the Outback

Wycliffe Well is in the middle of nowhere in Australia's Northern Territory. People there claimed to see many UFOs. It became the UFO Capital of Australia.

Since the 1800s, Wycliffe Well served as a watering hole. The road house there provided gas, food, and beds for Earthling travelers. During World War II, soldiers used Wycliffe Well as

The gas station at Wycliffe Well.

a stopover. Around that same time, the UFO sightings began.

Lew Farkas owned the road house in 1985. He used a diary to collect UFO stories from locals, indigenous tribes, and travelers. He turned the road house into an alien theme park. He believed an energy drew aliens to Wycliffe Well. During a month-long UFO frenzy, bright lights in the sky terrified Wycliffe Well guests. In the chaos, kangaroos escaped from the petting zoo.

Aerial view of Pine Gap.

1,000

Estimated number of members in Australia's many national UFO clubs

- Wycliffe Well is only four hours from a US spying facility called Pine Gap.
- The US military used the site to watch for enemy missiles.
- Pine Gap activities may have caused the many UFO sightings at Wycliffe Well.

12

UFO Welcome Center: We Come in Peace

Metal panels stretch over a wood frame. Junk litters the ground. Electrical wires dangle. Car jacks and stumps prop up the homemade spaceship that is the UFO Welcome Center in Bowman, South Carolina.

Jody Pendarvis.

Joseph "Jody" Pendarvis built the spaceship in 1994 to attract

The spaceship has a bed, bathroom, television, and air conditioning.

aliens to Earth. A highway billboard instructs UFOs to tilt right at the town of Bowman.

85,200

Number of times someone in South Carolina can look at the sky and never see a UFO

- South Carolina has had fewer than 6,000 UFO sightings ever.
- Chances of encountering any aliens are lowest in Florida.
- Pendarvis might want to move to South Dakota. It has the best odds in the US.

The welcome ship measures 46 feet (14 m) across. A second, smaller flying saucer balances above. Pendarvis climbs atop his ships and watches for UFOs. Years ago, he saw a UFO that looked like the moon. He hopes aliens will carry him away.

A guest book at the UFO Welcome Center asks visitors what planet they come from. So far, only earthlings have passed through.

Above and Beyond

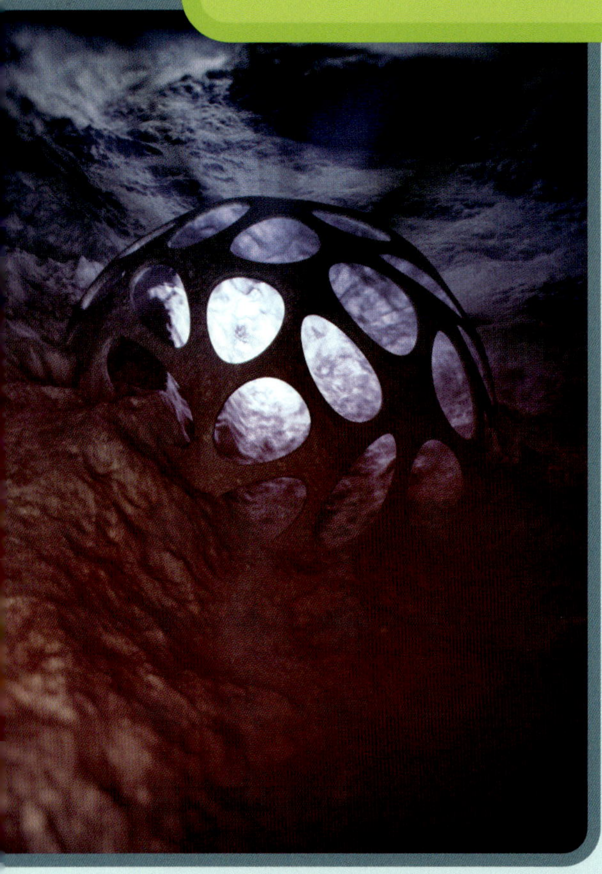

Unidentified Submerged Objects (USOs)

Strange glowing objects zip through the ocean, too. The Redondo Trench in the Pacific Ocean is a popular USO spot. Some believe that Christopher Columbus saw a bright object under the water in 1492. It flew away into the sky.

UFO Festival, Roswell, New Mexico

Aliens and alien fans can visit this annual UFO Festival. Famous ufologists give talks.

Music booms. There are parades and out-of-this-world costume contests.

Alien Vacations

Travelers to Joshua Tree, California, can book a flying saucer. The saucer is a Futuro House. Finnish architect Matti Suuronen created these houses in the 1960s and 1970s. In Wales, beings from all galaxies can stay in the Spodnic House. This metal pod was originally used in the 2012 London Olympics.

Glossary

conspiracy
An explanation of events involving sinister forces such as aliens or UFOs.

cosmos
The physical universe of all space and matter including planets, stars and galaxies.

debunk
To show that something is not true.

geoglyph
A large design created on the ground.

kiln
An oven used to dry and harden pottery.

lair
A secret hiding place where a creature can hide.

raid
A sudden attack by a small group to seize something.

stealth
A secret, quiet way of moving, or an aircraft that cannot be detected by radar.

telepathic
Having the ability to read minds or transmit thoughts.

UFO
Unidentified flying object.

Read More

Hunter, Nick. *Top Secret. Have Aliens Visited Earth?* Chicago, IL: Raintree. 2017.

Manzanero, Paula K. *Where Is Area 51?* New York: Penguin Workshop, 2018.

Tarakson, Stella. *Aliens, Ghosts and Vanishings: Strange and Possibly True Australian Stories.* Sydney, Australia: Penguin Random House Australia, 2016.

Visit 12StoryLibrary.com

Scan the code or use your school's login at **12StoryLibrary.com** for recent updates about this topic and a full digital version of this book. Enjoy free access to:

- **Digital ebook**
- **Breaking news updates**
- **Live content feeds**
- **Videos, interactive maps, and graphics**
- **Additional web resources**

Note to educators: Visit 12StoryLibrary.com/register to sign up for free premium website access. Enjoy live content plus a full digital version of every 12-Story Library book you own for every student at your school.

Index

About the Author

Jenny Mason has always lived in UFO country. She grew up a few hours from Roswell, New Mexico. She currently lives in Colorado near the UFO Highway. She will go anywhere in search of amazing stories.